I AM GOD, AND THERE IS NONE LIKE ME, DECLARING THE

I AM GOD, AND THERE IS NONE LIKE ME, DECLARING THE

END FROM THE BEGINNING AND FROM ANCIENT TIMES THINGS NOT YET DONE. *Isaiah 46;*

Thats quite a statement? Is it true? Lets take a look.

Isaiah is believed to have lived in the 8th century B.C., He is believed to have prophesied the birth and ministry of Jesus Christ, Lets start with one of his prophecies which is easily verifiable in history. Isaiah 44; 28. Cyrus Is my shepherd, he shall perform my work, even saying to Jerusalem you shall be built, and the temple and its foundations shall be laid.

Isaiah 45; Thus says the lord (God) to his anointed (chosen servant), CYRUS, whose right hand i have held, to subdue (conquer) nations, and I the lord will give you treasures, that you may know that I, the lord, who calls you by name, (naming Cyrus) am the lord of Israel. For jacob my servants sake, and Israel my chosen, i have even called you by your name, even though you dont know me. When did isaiah live?

Cyrus the Great (580-529 B.C.) the first Persian Emperor. He founded Persia by uniting the two original Iranian Tribes- the Medes and the Persians. Although he was known to be a great conqueror, who at one point controlled one of the greatest Empires ever seen, he is best remembered for his unprecedented tolerance and benevolent attitude towards those he defeated.

Upon his victory over the Medes, he started a government for his new kingdom, incorporating both Mede and Persian nobles as civil officials. The conquest of Asia Minor completed, he led his armies to the eastern frontiers. Hyrcania and Parthia were already part of the Median Kingdom. Further east, he conquered Drangiana, Arachosia, Margiana and Bactria. After crossing the Oxus, he reached the Jaxartes, where he built fortified towns with the object of defending the farthest frontier of his kingdom against nomadic tribes of Central Asia.

The victories to the east led him again to the west and led him to attack Babylon and Egypt. When he conquered Babylon, he did so to cheers from the Jewish Community, who welcomed him as a liberator- he allowed the Jews to return to the promised Land (after the babylonian captivity). He showed great respect towards the religious beliefs and cultural traditions of other races. These attributes earned him the respect and homage of all the people over whom he ruled.

Cyrus the Great, in Pasargad, in modern day Iran. The victory over Babylon expressed all the facets of the policy of conciliation which Cyrus had followed until then. He presented himself not as a conqueror, but as a liberator and the legitimate successor to the crown. He also declared the first charter of human rights known to mankind (see Cyrus's stellar). Cyrus had the wisdom to leave unchanged the institution of each kingdom he attached to the Persian Crown. In 539 B.C. he allowed more than 40,000 Jews to leave Babylon and return to Palestine (where they finished building the Jewish temple). This step was in line with his policy to bring peace to Mankind. Liberating nations from slavery.

Cyrus was upright, a great leader of men, generous and benevolent. The Hellenes, whom he conquered regarded him as 'Law-giver' and the Jews as 'the anointed of the Lord' (as Isaiah's prophecy stated).

Isaiah is recognised as being alive in the 8th century B.C. and Cyrus over 100 years later in the 6th century B.C. So if this is true then Cyrus was prophesied to rule the largest empire in the world at that time over 100 years before he was born.

Is this definite proof of the existence of God? No, there are alternative explanations. Isaiah might have lived at the same time as Cyrus, perhaps Cyrus's parents heard about the prophesy and named their child Cyrus to help him gain power, etc, etc. So did Cyrus come to prominence because God "declared it", or was it just by natural means. Cyrus is just one of many prophecies which "God declared" (I am God, declaring things from ancient times). So let's investigate the other prophecies.

The prophecy of the passover being kept and passed down through the ages.

THE ISRAELITES HAVE LEFT SLAVERY IN EGYPT AFTER GOD WON THE CONTEST AGAINST EGYPTS SO CALLED GODS, THE LAST CONTEST ENDING IN THE DEATHS OF THE EGYPTIANS FIRST BORN SONS, WHICH IS STILL CELEBRATED TODAY AS THE PASSOVER (THE DESTROYER PASSED OVER THE HOUSES OF THE ISRAELITES AND ONLY KILLED THE EGYPTIANS) BY THE JEWS AND SOME CHRISTIAN CHURCHES.-

EXODUS 12 - 14 ,

THIS DAY YOU SHALL KEEP AS A MEMORIAL,YOU SHALL KEEP THIS FEAST TO THE LORD THROUGHOUT YOUR GENERATIONS, YOU SHALL KEEP IT AS A FEAST FOREVER,

WHEN YOUR CHILDREN ASK YOU WHAT IS THE MEANING OF

THIS FEAST YOU SHALL SAY ,

"IT IS THE LORDS PASSOVER", WHO PASSED
OVER THE HOUSES OF THE CHILDREN OF ISRAEL IN EGYPT
WHEN HE KILLED THE EGYPTIANS.

3500 YEARS LATER ITS STILL BEING TOLD AND
CELEBRATED.

THE ISRAELITES LEAVE EGYPT AFTER BEING THERE FOR 430 YEARS.

So here once again, Gods prophecy is 100% true. The PASSOVER is being celebrated.

Deuteronomy 28/15;

But it shall come to pass if you will not listen to God and not do his commandments, then all these curses will come upon you and overtake you.

Deuteronomy 28/25;

God shall cause you to be smitten by your enemies, and you will be removed into all the kingdoms (countries) of the earth.

That is history, referred to as the great Jewish Diaspora, but the Jews only accounted for 2 tribes of the Israelites and are known because they have kept their identity for all this time, the other 10 tribes are referred to in history as the 10 lost tribes of Israel.

Deuteronomy 28/37;

And you shall become an astonishment, a proverb and a byword among all the nations where God shall lead you.(Anti semitism-hatred of the Jews). (Which has happened throughout history and is still with us today).

Deuteronomy 28/64;

And God shall scatter you among all people, from one end of the earth to the other. *28/65;* And among these nations you shall find no ease. *28/66;* And your life shall hang in doubt before you.

This is exactly what has happened in history. The Jews have been persecuted and **feared for their lives** throughout the world. After the Nazi holocaust of WW2 the Jews finally returned to Israel and set up their own state in 1948 to be able to determine their own fate. So far this is the **3rd indisputable prophecy** to have come to pass. 'What will happen over the next 25 years? The being who is referred to worry, this is only the start of many prophecies that have come to pass. Only **fools** say none of Gods prophecies have come to pass.

So what is the REAL MESSAGE of the Bible? Is it that some people will go to some vague "heaven" for being good and the others will go to some terrible "hell" where they will suffer unbelievable pain and suffering for all eternity? Is that what you believe? The real story is quite surprising, but the Bible states that GOD himself will come to planet Earth to live with his "children" 1000 years after Jesus Christ returns in the "Clouds of the Lord" and sets up his government for all of us except the incorrigibly wicked who will not leave their evil ways.

This is made clear by Gods statement that he doesn't want to lose even one of us, and after the "final judgement" the entire Universe will be placed under our "custodianship" similar to the Earth being currently under mankind's dominion. This is why God refers to himself as being our "Father" and JESUS CHRIST referring to himself as our brother, who loves us enough to suffer extreme agony and death to show us the way to immortality. Ultimately there are two ways with two differing ends.

One is Gods way which leads to eternal life with God and Jesus as our father and older brother, the other way which leads to eternal death. The Bible describes the dead being resurrected to life. Initially 144,002 at the return of Christ and 1000 years later the rest of all humanity who have ever lived. It all makes so much sense once properly understood. Don't be fooled, investigate yourself, eternal wonderful life without suffering, or eternal death with no suffering (euthanasia), these are the real choices and the only choices.

http://www.amazon.com/s/ref=nb_sb_noss?url=search-alias%3Ddigital-text&field-keywords=clouds%20of%20the%20lord%20old%20testament

What will happen over the next 25 years? The being who is referred to as Jesus Christ left a record of the events leading up to his return in "the Clouds of the Lord". Such events are recorded in "the Olivet Prophecy". Make no mistake Bible prophecy has always been fulfilled, and always will be. Do not be fooled by those who do not understand prophecy, "Clouds" of the Lord are one area of understanding that can truly open ones mind to the "real truth".

John 20/30; But these are written, so you might believe that jesus is the son of God. In other words prophecy. You can believe Christ was an immortal because he revealed the future.

When you understand Prophecy you understand Jesus was no ordinary mortal.

You understand the future before it happens and your faith is not "blind", it is extremely informed. Mathew 24/34 ; Verily i say unto you, this generation will not pass, till all these things be fullfilled. St luke 21/32; Verily i say unto you, this generation shall not pass away till all be fulfilled. Which generation?

Many atheists claim Jesus was talking about his generation, but Jesus clearly stated that "The Gospel must be preached to the entire world, then the end of this age will come. Only in this generation is that being fulfilled, the Gospel (meaning good news) is finally being preached to the whole world with the use of the internet and satellite tv.

Other prophecies from Deuteronomy to Revelation state the Jews and other Israelites would be dispersed throughout the entire world and then brought back to Israel in the "last days".

This happened with the establishment of the nation of Israel again in 1948 after the Nazi holocaust. Other Prophecies say that Christ will come back in the last day's just before mankind destroys everything.

This only became possible in the 50's when the U.S. and the Soviet Union built up nuclear arsenals capable of killing everyone. How did they know all this would occur within one generation?

Quite simple, the message was from the creator, God.

In The Bible God states "I am God, and there is none like me, declaring from ancient times things that will come to pass". Did these things come to pass? 100% they have, and more is coming.

Many leading atheists claim nothing the Bible states as Prophecy has come to pass, and when some points are shown, (such as the Jews returning to Israel) they claim they are self fulfilled, ie; They made it happen (The Jews). The information on the next page will show you how ridiculous this statement is, once the Bible is studied extensively and humbly you can see the arrogant self inflated atheist views are extremely stupid and simplistic, just as the Bible prophecied they would be 2 peter 3; Knowing this first, that there shall come in the last days scoffers, walking after their own lusts (perverted sexual behaviour, not natural,such as porn and related filth), And saying, Where is the promise of his coming? for since the fathers fell asleep, all things continue as they were from the beginning of the creation . **It is easy to see that many "reject"God and the Bible for their own "Moral" reasons, not logical ones, others are merely indoctrinated into atheist or false religious views.**

There is one word you will hear from atheists, whether they have little education or they are oxford or harvard "trained" (brainwashed)educators. That word is "coincidence", used over and over again to explain extreme levels of design.

One that comes to mind is the Moon, which appears to be about the same size as the Sun in the sky even though the Moon is tiny in comparison, the reason being it is at a relative distance to the Earth. It is also at just the right distance to allow human life to exist on earth.

I remember an atheist scientist saying, "owing to an amazing coincidence", the Moon, even though it is four hundred times smaller than the Sun, it is four hundred times closer to the Earth and that is the reason it appears to be the same size in our sky. "Coincidence"?, "Amazing coincidence"?, or is this more likely?

Genesis 16

And God made two great lights, the greater light (the Sun) to rule the day, and the lesser light to rule the night (the Moon).

Beings able to travel from someplace perhaps billions of light years away or maybe not even in our universe would certainly be able to move the moon into a desired orbit around the Earth.

We launch satellites and place them into our desired orbits around the planet and we haven't even walked on Mars yet.

Beings who also tell us about a war, starting in Iraq

(the Iraq war has unleashed massive hostility towards the West and unbalanced the balance of power in that entire region and unleashed islamic "jihad" with shi-ite and sunni jihadists unleashing a campaign of terror bombings upon each other and the rest of the world),

which will ultimately kill one third of mankind, in other words a World War surpassing the world wars we have seen so far, WW1- 9 million dead, WW2- 70 million dead, WW3- ? dead, one third of the earths population.

It has already started, at the end of the Second World War U.S. army general Douglas Macarthur (1880-1964) stated "<u>we have had our last chance</u>, war has become so destructive, so dreadfull, if we do not find an alternative, we will destroy civilisation.

He was at that time in charge of the allied occupation of Japan after their surrender to the Allies, which only came about after the U.S. detonated nuclear bombs over the Japanese cities of Hiroshima and Nagasaki (the Manhatten project) totally destroying them and threatening to destroy the rest of Japan with more nukes because of the suicidal and homocidal fanaticism of the Japanese (today is no different, we have fanatics who launch suicidal attacks across the planet, what will happen when these fanatics get weapons of mass destruction?).

After the first world war the League of Nations was formed to prevent another world war, they utterly failed with the second world war 20 years later with about 10 times more destruction.

The United Nations, the successor to the League of Nations has not even been able to prevent or stop many small wars (Rwanda, Congo, Sierra Leone and dozens of others) and is not stopping WW3.

When Albert Einstein died on April 18, 1955 he left a piece of writing ending in an unfinished sentence (a warning).

These were his last words:

In essence, the conflict that exists today is no more than an old-style struggle for power, once again presented to mankind in semi religious trappings. The difference is that, this time, the development of <u>atomic power</u> has imbued the struggle with a ghostly character; for both parties know and admit that, should the quarrel deteriorate into actual war, <u>mankind is doomed</u>.

Despite this knowledge, statesmen in responsible positions on both sides continue to employ the well-known technique of seeking to intimidate and demoralize the opponent by marshaling superior military strength. They do so even though such a policy entails the risk of war and doom. Not one statesman in a position of responsibility has dared to pursue the only course that holds out any promise of peace, the course of supranational security, since for a statesman to follow such a course would be tantamount to political suicide. Political passions, once they have been fanned into flame, exact their victims.

Here we have warnings from two of the most influential men of the preceeding generation about our future, even without the prophecies of the Bible it is plain to see the human race is headed for real trouble. With an accurate knowledge of the Bible's (God's) prophecies it is possible to understand what is going to happen years before it actually happens. With an inaccurate understanding of the prophecies obviously an inaccurate understanding of events will occur. This has been happening for centuries, and ultimately will still occur till Christ returns in the "Clouds of the Lord".

We have enough weapons of mass destruction today to destroy all life with many more countries obtaining these weapons of mass destruction over the next decade. Following the pattern of the previous world wars the death rate is likely to be much higher.

Unless someone or something intervenes it would likely be much higher than WW1 or WW2. 60% 70% 80% 90% or even 100%, complete genocide. We need a saviour, let's hope one will come,

"IN THE CLOUDS OF THE LORD".

What did these beings say? Did it occur?

http://www.amazon.com/s/ref=nb_sb_noss?url=search-alias%3Ddigital-text&field-keywords=clouds%20of%20the%20lord%20old%20testament

Did these beings relay anything of importance? Are their revelations backed up by history or science?

Lets have a look at some of the plain and simple revelations and see if they occured independently from the Bible or "religious" sources.

The Bible says in *Isaiah 46;9, "Remember the former things of old: For i am God, and there is none else, I am God and there is none like me". Isaiah 46;10;Declaring the end from the beginning, and from ancient times the things that are not yet done"* (prophecy, saying *my counsel shall stand, and i will do all my pleasure,* (in other words whatever God declares will happen, will happen regardless of what any one says, even if it was prophesied by God in ancient times it will "come to pass". Has it?

Most people are aware of the Jews returning to form the modern day state of Israel after the holocaust in WW2 in 1948. Did they say anything about that?

Deuteronomy 28;

And it will happen, if you will listen to God and obey all his commandments. which i command you this day, that God will set you high above all nations of the world.

Moses tells the Israelites in the desert what God has stated, that if they obey him they will be the worlds leading nations. and that everything will go well for them.

Deuteronomy 28/15;

But it shall come to pass if you will not listen to God and not do his commandments, then all these curses will come upon you and overtake you.

Deuteronomy 28/25;

God shall cause you to be smitten by your enemies, and you will be removed into all the kingdoms(countries)of the earth.

That is history, referred to as the great Jewish Diaspora, but the Jews only accounted for 2 tribes of the Israelites and are known because they have kept their identity for all this time *(Napoleon commented on how the Jews kept their identity with such tenacity)*, the other 10 tribes are referred to in history as the 10 lost tribes of Israel.

Deuteronomy 28/37;

And you shall become an astonishment, a proverb and a byword among all the nations where God shall lead you.(Anti semitism-hatred of the Jews). (Which hashappened throughout history and is still with us today).

Deuteronomy 28/64;

And God shall scatter you among all people, from one end of the earth to the other. 28/65; And among these nations you shall find no ease. 28/66; And your life shall hang in doubt before you.

This is exactly what has happened in history. The Jews have been persecuted and feared for their lives throughout the world throughout history. After the Nazi holocaust of WW2 the Jews finally returned to Israel and set up their own state in 1948 to be able to determine their own fate without being persecuted.

Deuteronomy 30;

And it shall come to pass when all these things have come upon you, the blessing and the curse, which i have set before you and you shall remember them in all the nations God has driven you to.

And return to the Lord, your God, and you obey him again according to all that I command you this day, you and your children, with all your heart and all your soul.

Then the Lord your God will turn your captivity, and have compassion on you, and will return and gather you from all the nations, where the Lord your God has scattered you.

If any of you has been driven out to the outermost part of the earth the Lord God will gather you from there, and will bring you back into this land, which your fathers posessed, and you shall posess it and he will do you good and multiply you more than your fathers.

This is exactly what has happened, a homeless, wandering group of tribes forming an ancient nation, turning against God, ultimately losing their nation, being dispersed throughout the world and then after turning back to God during and after the holocaust of WW2 they formed the modern nation of Israel.

Atheists say they returned because the Jews knew about the prophecy that they would return or even that the Jews of today are not related to the Jews of the times of the ancient Israelites, the deception in the world is that profound, people even believe the Jews murder Christian(Catholic) children for their blood to make their passover bread.

Would they intentionally spread to all parts of the world and be hated (Anti-Semitism) being killed, tortured, persecuted, executed, etc;? No, that is a ridiculous assumption put forward by militant atheists and religious fanatics to deny God or justify murdering the Jews. And there is much more.

Deuteronomy 31/15; "

I have set before you this day "life and good, and death and evil' 31/19;-I call heaven and earth to record this day against you, that i have set before you life and death, blessing and cursing, therefore choose life, that you and your descendents may live.

The Jews have been persecuted throughout history as foretold in the Bible. The Assyrians conquered and deported the northern ten tribes of Israel to the land of the Medes in modern day Iran, and the Babylonians conquered the southern 2 tribes (the Jews) and took them to Babylon (modern day Iraq). The Romans enslaved the Jews and ultimately banned them from Jerusalem on pain of death in 136 A.D. With huge numbers being killed and taken to Rome as slaves.

The Roman Catholic Church persecuted the Jews throughout the ages stating the Jews were the murderers of Christ with Popes issuing decrees against Jews and inciting hatred.

The Greek Orthodox Church also instigated "pogroms" against them with many, many being tortured and killed. Islam has also killed and persecuted the Jews throughout the ages and now in modern times have exiled them from many Islamic nations and regularly attack them swearing to annihilate them and finish off Hitlers "final solution of the Jewish problem" (see Grand Mufti of Jerusalem issuing fatwa against the jews in the 1920's, he became Hitlers ally in 1939).

Communist Russia imprisoned the Jews and persecuted them (eventually allowing thousands to migrate to Israel in the 1980's). Even this tiny amount of information should show you the accuracy of this unlikely prophecy. Much more will be revealed on this site. Keep a humble and open mind and prove all things, dont believe all things (what you are told to believe or what you want to believe). Contact me if you have questions and i will endeavour to answer them honestly, but i will only answer humble people, arrogant people can believe what they choose to believe.

Read carefully the prophecies mentioned at the top of this page (the Jews being an astonishment, a proverb and a byword-anti-semitism) and the Jews life hanging in the balance, and the Israelites being dispersed throughout the world, and being brought back to the ancient land of Israel. There is much, much more that will be revealed here over the years while the author is alive. I apologise for being slow at doing this but hopefully the accuracy of what is written here will make up for that.

135 B.C.;The Syrian Greek king Antiochus Epiphanes desecrates the second Jewish temple (built by the remnant of Jews who returned under the Babylonian and Persian captivity) causing the Hasmonean revolt against the Greek authorities.

70 A.D.; Titus of Rome takes Jerusalem - killing 1 million Jews.(prophesied by Jesus=see Jesus wept).

136 A.D.; Third Jewish revolt- Jews killed and banned from living in Jerusalem under pain of death (renamed Aelio Capitolina by the Romans).

306 A.D.; Council in Spain bans Christians and Jews marrying.

1012 A.D.; Emperor Henry of Germany expels Jews from Mainz, beginning German persecution (over 900 years before Hitler).

1096 A.D. ; 1st crusade - Crusaders massacre Jews in the Rhineland.

1144 A.D.; First recorded case of "Jewish blood libel" where anti- Jewish elements claim the Jews use Christian children's blood for their Passover bread. (The Passover was commanded by God to be observed down through the ages-obviously without killing children), Slitting the child's throat and draining their blood to make the bread (the Islamic propaganda machine is claiming now that this is real, with millions of Muslims believing it today.)

1190 A.D.; Massacre of Jews in England.

1290 A.D.; Jews exiled from England.

1298 A.D.; Jews murdered in Germany.

1306 A.D.; Jews exiled from France.

1348 A.D.; Jews blamed for the "black death" (bubonic plague- the Jews did not suffer as much from the plague because they followed the instructions from God in Leviticus concerning cleanliness, disease and isolating the sick from the population causing them to be accused of being in league with the "devil" from the Catholic and Orthodox hierarchy.

1389 A.D.; Massacres in Bohemia -Spain.

1421 A.D.; Hundreds of Jews burned at the stake by the inquisition by the Catholic religious authorities.

(France already burnt many Jews at the stake by this time).

1480 A.D.; More Jews burnt at the stake in Spain.

1483 A.D. Exiled from Poland, Lithuania,Sicily and Portugal.

1492 A.D. All Jews exiled from Spain' (By Isabella and King Ferdinand).

1510 Jews exiled from Brandenburg Germany.

1569 Jews exiled from "Papal states" by the Pope.

1593 Jews exiled from Bavaria and Italy.

1598 Jews executed for "blood libel", accused of killing Christian children " for passover bread", executed by being "quartered", ripped into quarters.

1648 100,000 Jews killed in Ukraine by the Cossacks.

1715 Pope Pius 6th orders edict against the Jews.

1768 Jews massacred in Poland. 20,000

1805 Jews massacred in Algeria (so much for peacefull Muslims).

1840 Blood libel again in Damascus Syria against the Jews.

1853 Blood libel by the Orthodox church in Russia.

1881 Pogroms (killing, raping 100,000 Jews in Kiev Ukraine.

1919 3000 Jews killed in hungarian pogroms.

1929 Hebron- Muslims massacre hundreds of Jews, cut heads off Jewish children and stab their mothers to death (see British police chiefs testimony "Ray Cafferata's testimony to the british court").

1939- 1945 Massacre of millions of Jews in "death camps" by the Nazis.

1948 Murder and exile of Jews across the Muslim world.

1948 - To present, many, many attacks by Muslims against Israel and Jews across the world, with Muslims swearing to "slaughter the Jews" and "drink the blood of the Jews".

Zechariah chapter 12; - God says he will make Jerusalem a *"cup of trembling" to the people round about it and the nations surounding it shall lay siege against it.* Precicely what is happening today.

The Islamic countries surrounding Jerusalem and Israel are involved in planning its destruction and conversion to Islam. The Bible describes what is currently happening in the book of Psalms, written in king Davids time 3000 years ago. *Psalm 83;3; "They* (the Islamic countries of Egypt, Iran, Iraq, Lebanon, Jordan and Turkey) *have taken crafty counsel against your people* (Israel). *Psalm 83;4; They* (The Islamic countries) *have said, come and let us stop them* (the Israeli's) *from becoming a nation, that the nation of Israel will be no more.*

Psalm 83;5; For they (the Islamic countries), *have consulted together with one mind, they are working together against Israel.* This is exactly what has been happening since the Jews returned to Israel in 1948 and even before that. The Muslim Brotherhood in Egypt (a worldwide Sunni islamist organisation) with its offspring Hamas in Gaza which has constantly been attacking Israel, the Shi-ite Mullahs of IRAN with their Shi-ite offspring Hezbollah in Lebanon (responsible for killing 100's of u.s. soldiers in lebanon and having a war against Israel) have all sworn the destruction of the Jewish state of Israel and call it a cancer which must be destroyed, just as God prophesied in the Bible 3000 years ago.

When you have a proper understanding of the Bible you start to get an understanding of what is going to happen because as the Bible says, "God declares what is going to happen from ancient times, and he has the power to either know it is going to happen or has the ability to make it happen. The history of mankind has been pre recorded to some extent to show those willing to follow God what will happen, and you can be sure that whatever God declares will happen, will happen, regardless of whatever any human being says.

Jeremiah 30/1; 600b.c. The word that came to Jeremiah from the Lord, saying, thus speaks the lord of Israel, saying, write what i have told you in a book (much the same as Moses had to write in a book, also the apostle John was commanded to write in a book).

For the days are coming when i will bring Israel and Judah into captivity, fear not i will save you from afar, from the land of their captivity.

And Jacob (The Jews) *shall return, though i will make an end of all nations where i scatter you, I will not make a full end of you* (all the Empires have passed into history, the Assyrians, the Babylonians, the Persians, the Greeks, the Romans, the Ottomans, the Nazi's). But the Jews have been preserved, unbelievable but true.

But i will correct you in measure and will not leave you unpunished. (The Jews have certainly been punished). *Therefore all those that devour you shall be devoured,* once again the nations which "devoured" the Jews have themselves been "Devoured" by other nations. *Jeremiah 33/7; I will cause the captive Jews to return and the captive Israelites.* The Jews have returned but the other Israelites have not yet. *Jeremiah 34/17; I will make you to be removed to all the kingdoms of the earth.* Did this happen? Very much so.

Again I will build you and you will be built. You shall plant vines in Samaria, (The west bank). *Behold i will bring them from the north country* (Europe) *and gather them from the coasts of the earth, hear the word of the Lord o' nations he that scattered Israel shall gather him. They shall come and sing in Zion* (Jerusalem), *then the virgin shall dance* (look at the video's of the young women dancing at the declaration of Israel in 1948).

The prophecy of Babylon being destroyed and not being rebuilt, has it? Jerusalem and many other cities are still with us, but Babylon? no international flights to Babylon international airport is there? why? because God decreed it thousands of years ago.

CLOUDS OF THE LORD-NEW TESTAMENT

http://www.amazon.com/s/ref=nb_sb_noss?url=search-alias%3Ddigital-text&field-keywords=clouds%20of%20the%20lord%20old%20testament

FORWARD

The New Testament, which confirms and continues on from the Old Testament, primarily around events from just before the birth of Jesus Christ, through to the ministry, death (by crucifixion) and "alleged" resurrection three days later, through to the early days of the seven churches, on through history to the present time and into the future.

The continuity of the involvement of the Clouds of the Lord in the New Testament is absolutely astounding. Jesus Christ "coming with Clouds"? What are these "Clouds"?

The ancient writers of the books of the Bible, all who claim they had contact with God by Angels or visions, in many different time periods, have inadvertantly revealed interactions with beings and (for want of a better word) vehicles far superior to anything in our worlds level of current technology.

Their descriptions within their own limited scope of learning, naming these vehicles as "Clouds of the Lord" and "Chariots of fire" and "Lights from Heaven".

Inadvertantly revealing vehicles of extreme levels of technology, which is to be expected from beings who are able to visit Earth from anywhere else in this universe owing to the extreme distances and hazards of such a journey from outer space, or if they come from or through another dimension.

Their description in events past, present and future fit into a pattern of logic that is well beyond the many writers from many different ancient time periods to collude on and falsify.

So much so that any rational, logical thinking person today with any reasonable amount of intelligence would have to concede is based on facts. In other words if you are intelligent enough to read this page and understand what is written on this page, you have enough intelligence to understand the rest of this book. You dont have to be a university professor at oxford or harvard. You only need to slightly open your mind and to have a tiny amount of humbleness. What is the purpose of these beings who visit our planet in these "Clouds" and "Chariots"? According to these ancient writers, all summed up together, the answer will impact on every human being who has ever lived. Read on and make up your own mind.

Introduction

http://www.amazon.com/s/ref=nb_sb_noss?url=search-alias%3Ddigital-text&field-keywords=clouds%20of%20the%20lord%20old%20testament

The first mention of one of these incredible "vehicles" describes it as a "Star" in the book of Mathew. Mathew describes it from the revelation of the three "wise men from the east" who are looking for the messiah *(Jesus Christ)*.

It appeared to look like a star *(known now as the "Star" of Bethlehem to the world)* because it was high up in the earths atmosphere, as opposed to it looking like a "Cloud" or "Fire" or "Chariot of Fire" when closer to the observing witness in other parts of the Bible. In the book of revelation John reports seeing an Angel "clothed with a Cloud" coming down to Earth in a vision given to him from Jesus Christ many years after Jesus crucifixion.

The ancient Israelites followed a "Cloud" for forty years in Moses time, Moses went into a "Cloud" and got the ten commandments from God. Similar accounts are in the book of Luke.

The book reports an angel of the Lord coming to shepherds at night and the "Glory of the Lord" shining around them, they being obviously terrified the Angel responds to their terror, fear not, and states the prophesied Messiah has been born *(Jesus Christ)*.

The Angel's vehicle appears to look like a star to these men much like a helicopter or plane at night high up in the sky would appear to primitive men in today's world. It's quite easy to mistake a helicopter or plane for a star temporarily today.

MATHEW 2- Now when Jesus was born in Bethlehem, in king Herods day, wise men from the east came to Jerusalem asking "where is he that is born king of the Jews"? for we have seen his "Star" in the east and have come to worship him. King Herod upon hearing this was troubled, *(fearing for his kingship, not being a Jew)* **and he asks the Jewish priests where Christ would be born. They reply in Bethlehem because it is written in the books of the Prophets. Then Herod calls the wise men and asks them when the "Star" appeared and he sends the wise men to Bethlehem to find the child for him, telling them he wants to worship the child also** *(but planning to murder the child thinking it is a danger to his rule).*

The wise men leave king Herod and the "Star" appears, they follow it to Bethlehem and it stops above where the child is *(obviously not a Star like our Sun),* **much like Moses and the Israelites 1400 years previously when they followed the "Cloud" by day and the "Fire" by night** *(the "energy vehicle" being closer to the ground).*

They rejoiced with "exceeding great joy". They go in to see the child and present gifts to the family then leave and return to their homeland *(in the east)* **without telling King Herod where the child is.**

King Herod finds out and responds by slaughtering the children in the area *(also prophesied in the Old Testament)* **under the age of two.**

An Angel *(Energy Being)* **appears to the child's earthly father (Joseph) and warns him to flee to Egypt as king Herod wants the child Jesus dead because the prophecies describe Jesus as king of the Jews and coming to rule. Herod being fearful of his kingship.**

Joseph and his family stay in Egypt till the kings death then returns to his homeland.

As a grown man Jesus comes to John the Baptist *(also prophesied in the Old Testament)* in the wilderness *(John was actually Jesus cousin)*, Mathew 3;13 Then Jesus came to Jordan unto John to be baptised by him A.D.27.

And Jesus, when he was baptised, when out of the water, the Heavens were opened, and he saw the "Spirit of God" descending like a Dove, and lighting upon him, and a voice coming from Heaven, saying, this is my beloved Son in whom I am well pleased.

The next mention of one of these "vehicles" is five years later

Mathew 17; 32A.D.

And after six days Jesus took Peter, James and John his brother, and brought them up into a high Moutain, and Jesus was transfigured before them, and his face did shine *(similar to when Moses returned from being in the "Cloud" of the Lord for forty days on Mount Sinai 1400 years previously)* as the sun, and his raiment was white as the light, and behold, there appeared unto them Moses and Elijah talking with Jesus *(Moses and Elijah died 1400 and 900 years previously)* like old friends about Jesus approaching death in Jerusalem. Then Peter said to Jesus, Lord, it is good for us to be here, if you will, let us make three tabernacles *(tents)*, one for you, one for Moses and one for Elijah.

While he spoke, a bright "Cloud" *(energy vehicle)* overshadowed them, and a voice came out of the "Cloud" *(similar to the events of Moses time when God spoke to the Ancient Israelites from out of the "Cloud")*, which said, this is my beloved son, in whom I am well pleased, hear *(listen to)* him.

And when the disciples heard it, they fell on their faces, and were afraid *(understandably so)*. Jesus came and touched them and said, arise, and dont be afraid. When they looked they saw no man but Jesus. Moses and Elijah gone with the "Cloud" *(energy vehicle)*.

The next mention is in Mathew 24; 33A.D. The Olivet discourse.

Jesus is upon the Mount of Olives near Jerusalem talking to his disciples about his return at the end of the age and the world situation at that time.

"Immediately after the tribulation of those days shall the sun be darkened and the moon shall not give it's light, and the stars shall fall from heaven (meteorites), and the power of the heavens shall be shaken.

And then shall appear the sign of the son of man (Jesus Christ) in heaven, and then shall all the tribes of the earth mourn, and they shall see the son of man (Jesus Christ) coming in the "Clouds of Heaven" (the same way he left the Earth) with power and great glory.

These "Clouds" represent great power, much, much more than anything we have on Earth. The misguided armies of the Earth will attack the returning Jesus.

And he (Jesus) will send his Angels (energy beings) with the sound of a "trumpet", and they will gather together his "elect"(human beings chosen from throughout the ages such as the disciples numbering 144,002) from the four winds, from one end of heaven to the other.

Reinforced in Paul's letter to the church in Thessalonica (a Greek city) decades later. A.D.54, twenty years after Jesus death and ressurection chapter 3;13

But i would not have you ignorant, brethren, concerning them which are "asleep" (dead) that you sorrow not, the same way others do who have no hope (atheists).

For if we believe that Jesus died and rose again, even so, then those also who sleep in Jesus (chosen by Jesus who died), will God bring with him, for this we say unto you by the word of the Lord, that we who remainalive unto the coming of the Lord shall by no means precede those which are asleep (dead).

For the Lord himself shall descend from heaven with a shout, with the voice of an Archangel,and with the trump of God, and the dead chosen by Christ shall rise first.

Then we who are alive and remain shall be raised up together with them in the "Clouds" (they shall be transported into these "energy vehicles" which already contain the "Angels" and the ressureted "Elect" chosen by Jesus Christ, 144,002 individuals), to meet the Lord in the air (sky), and so shall we be always with the Lord.

Living and dead people from all over the world, from many different time periods, being transformed into "spirit" (energy beings-immortals) and being transported up into these energy vehicles "Clouds", where Jesus and his Angels (energy beings) already are in preparation to take over governing the planet just before the governments of the world destroy it.

Jesus left this Earth after his death and ressurection in one of these "Energy vehicles" and is returning with many of them according to these ancient writers.

Introduction

http://www.amazon.com/s/ref=nb_sb_noss?url=search-alias%3Ddigital-text&field-keywords=clouds%20of%20the%20lord%20old%20testament

The first mention of one of these incredible "vehicles" describes it as a "Star" in the book of Mathew. Mathew describes it from the revelation of the three "wise men from the east" who are looking for the messiah *(Jesus Christ)*.

It appeared to look like a star *(known now as the "Star" of Bethlehem to the world)* because it was high up in the earths atmosphere, as opposed to it looking like a "Cloud" or "Fire" or "Chariot of Fire" when closer to the observing witness in other parts of the Bible. In the book of revelation John reports seeing an Angel "clothed with a Cloud" coming down to Earth in a vision given to him from Jesus Christ many years after Jesus crucifixion.

The ancient Israelites followed a "Cloud" for forty years in Moses time, Moses went into a "Cloud" and got the ten commandments from God. Similar accounts are in the book of Luke.

The book reports an angel of the Lord coming to shepherds at night and the "Glory of the Lord" shining around them, they being obviously terrified the Angel responds to their terror, fear not, and states the prophesied Messiah has been born *(Jesus Christ)*.

The Angel's vehicle appears to look like a star to these men much like a helicopter or plane at night high up in the sky would appear to primitive men in today's world. It's quite easy to mistake a helicopter or plane for a star temporarily today.

MATHEW 2- Now when Jesus was born in Bethlehem, in king Herods day, wise men from the east came to Jerusalem asking "where is he that is born king of the Jews"? for we have seen his "Star" in the east and have come to worship him. King Herod upon hearing this was troubled, *(fearing for his kingship, not being a Jew)* and he asks the Jewish priests where Christ would be born. They reply in Bethlehem because it is written in the books of the Prophets. Then Herod calls the wise men and asks them when the "Star" appeared and he sends the wise men to Bethlehem to find the child for him, telling them he wants to worship the child also *(but planning to murder the child thinking it is a danger to his rule)*.

The wise men leave king Herod and the "Star" appears, they follow it to Bethlehem and it stops above where the child is *(obviously not a Star like our Sun)*, much like Moses and the Israelites 1400 years previously when they followed the "Cloud" by day and the "Fire" by night *(the "energy vehicle" being closer to the ground)*.

They rejoiced with "exceeding great joy". They go in to see the child and present gifts to the family then leave and return to their homeland *(in the east)* without telling King Herod where the child is.

King Herod finds out and responds by slaughtering the children in the area *(also prophesied in the Old Testament)* under the age of two.

An Angel *(Energy Being)* appears to the child's earthly father *(Joseph)* and warns him to flee to Egypt as king Herod wants the child Jesus dead because the prophecies describe Jesus as king of the Jews and coming to rule. Herod being fearful of his kingship.

Joseph and his family stay in Egypt till the kings death then returns to his homeland.

As a grown man Jesus comes to John the Baptist *(also prophesied in the Old Testament)* in the wilderness *(John was actually Jesus cousin),* Mathew 3;13 Then Jesus came to Jordan unto John to be baptised by him A.D.27.

And Jesus, when he was baptised, when out of the water, the Heavens were opened, and he saw the "Spirit of God" descending like a Dove, and lighting upon him, and a voice coming from Heaven, saying, this is my beloved Son in whom I am well pleased.

The next mention of one of these "vehicles" is five years later

Mathew 17; 32A.D.

And after six days Jesus took Peter, James and John his brother, and brought them up into a high Moutain, and Jesus was transfigured before them, and his face did shine *(similar to when Moses returned from being in the "Cloud" of the Lord for forty days on Mount Sinai 1400 years previously)* as the sun, and his raiment was white as the light, and behold, there appeared unto them Moses and Elijah talking with Jesus *(Moses and Elijah died 1400 and 900 years previously)* like old friends about Jesus approaching death in Jerusalem. Then Peter said to Jesus, Lord, it is good for us to be here, if you will, let us make three tabernacles *(tents)*, one for you, one for Moses and one for Elijah.

While he spoke, a bright "Cloud" *(energy vehicle)* overshadowed them, and a voice came out of the "Cloud" *(similar to the events of Moses time when God spoke to the Ancient Israelites from out of the "Cloud")*, which said, this is my beloved son, in whom I am well pleased, hear *(listen to)* him.

And when the disciples heard it, they fell on their faces, and were afraid *(understandably so)*. Jesus came and touched them and said, arise, and dont be afraid. When they looked they saw no man but Jesus. Moses and Elijah gone with the "Cloud" *(energy vehicle)*.

The next mention is in Mathew 24; 33A.D. The Olivet discourse.

Jesus is upon the Mount of Olives near Jerusalem talking to his disciples about his return at the end of the age and the world situation at that time.

"Immediately after the tribulation of those days shall the sun be darkened and the moon shall not give it's light, and the stars shall fall from heaven (meteorites), and the power of the heavens shall be shaken.

And then shall appear the sign of the son of man (Jesus Christ) in heaven, and then shall all the tribes of the earth mourn, and they shall see the son of man (Jesus Christ) coming in the "Clouds of Heaven" (the same way he left the Earth) with power and great glory.

These "Clouds" represent great power, much, much more than anything we have on Earth. The misguided armies of the Earth will attack the returning Jesus.

And he (Jesus) will send his Angels (energy beings) with the sound of a "trumpet", and they will gather together his "elect"(human beings chosen from throughout the ages such as the disciples numbering 144,002) from the four winds, from one end of heaven to the other.

Reinforced in Paul's letter to the church in Thessalonica (a Greek city) decades later. A.D.54, twenty years after Jesus death and ressurection chapter 3;13

But i would not have you ignorant, brethren, concerning them which are "asleep" (dead) that you sorrow not, the same way others do who have no hope (atheists).

For if we believe that Jesus died and rose again, even so, then those also who sleep in Jesus (chosen by Jesus who died), will God bring with him, for this we say unto you by the word of the Lord, that we who remainalive unto the coming of the Lord shall by no means precede those which are asleep (dead).

For the Lord himself shall descend from heaven with a shout, with the voice of an Archangel,and with the trump of God, and the dead chosen by Christ shall rise first.

Then we who are alive and remain shall be raised up together with them in the "Clouds" (they shall be transported into these "energy vehicles" which already contain the "Angels" and the ressurected "Elect" chosen by Jesus Christ, 144,002 individuals), to meet the Lord in the air (sky), and so shall we be always with the Lord.

Living and dead people from all over the world, from many different time periods, being transformed into "spirit" (energy beings- immortals) and being transported up into these energy vehicles "Clouds", where Jesus and his Angels (energy beings) already are in preparation to take over governing the planet just before the governments of the world destroy it.

Jesus left this Earth after his death and ressurection in one of these "Energy vehicles" and is returning with many of them according to these ancient writers.

BOOK OF MATHEW

http://www.amazon.com/s/ref=nb_sb_noss?url=search-alias%3Ddigital-text&field-keywords=clouds%20of%20the%20lord%20old%20testament

MATHEW 2- Now when Jesus was born in Bethlehem, in king Herods day, wise men from the east came to Jerusalem asking "where is he that is born king of the Jews"? for we have seen his "Star" in the east and have come to worship him. King Herod upon hearing this was troubled, *(fearing for his kingship, not being a Jew)* and he asks the Jewish priests where Christ would be born. They reply in Bethlehem because it is written in the books of the Prophets. Then Herod calls the wise men and asks them when the "Star" appeared and he sends the wise men to Bethlehem to find the child for him, telling them he wants to worship the child also *(but planning to murder the child thinking it is a danger to his rule).* The wise men leave king Herod and the "Star" appears, they follow it to Bethlehem and it stops above where the child is *(obviously not a Star like our Sun, many people mistakenly look for an actual "star" that might have "glowed" in Jesus time),* much like Moses and the Israelites 1400 years previously when they followed the "Cloud" by day and the "Fire" by night *(the "energy vehicle" being closer to the* *ground).* They rejoiced with "exceeding great joy". They go in to see the child and present gifts to the family then leave and return to their homeland *(in the east)* without telling King Herod where the child is. King Herod finds out and responds by slaughtering the children in the area *(also prophesied in the Old Testament)* under the age of two. An Angel *(Energy Being)* appears to the child's earthly father *(Joseph)* and warns him to flee to Egypt as king Herod wants the child Jesus dead because the prophecies describe Jesus as king of the Jews and coming to rule. Herod being fearful of his kingship. Joseph and his family stay in Egypt till the kings death then returns to his homeland. As a grown man Jesus comes to John the Baptist *(also prophesied in the Old Testament)* in the wilderness *(John was actually Jesus cousin),* Mathew 3;13 Then Jesus came to Jordan unto John to be baptised by him A.D.27.And Jesus, when he was baptised, when out of the water, the Heavens were opened, and he saw the "Spirit of God" descending like a Dove, and lighting upon him, and a voice coming from Heaven, saying, *this is mybeloved Son in whom I am well pleased.* The next mention of one of these "vehicles" is five years later, Mathew 17; 32A.D. And after six days Jesus took Peter, James and John his brother, and brought them up into a high Moutain, and Jesus was transfigured before them, and his face did shine *(similar to when Moses returned from being in the "Cloud" of the Lord for forty days on Mount Sinai 1400 years previously)* as the sun, and his raiment was white as the light, and behold, there appeared unto them Moses and Elijah talking with Jesus *(Moses and Elijah died 1400 and 900 years previously)* like old friends about Jesus approaching death in Jerusalem. Then Peter said to Jesus, Lord, it is good for us to be here, if you will, let us make three tabernacles *(tents),* one for you, one for Moses and one for Elijah. While he spoke, a bright "Cloud" *(energy vehicle)* overshadowed them, and a voice came out of the "Cloud" *(similar to the events of Moses time when God spoke to the Ancient Israelites from out of the "Cloud"),* which said, *this is my beloved son, in whom I am well pleased, hear (listen to) him.* And when the disciples heard it, they fell on their faces, and were afraid *(understandably so).* Jesus came and touched them and said, arise, and don't be afraid. When they looked they saw no man but Jesus.

Moses and Elijah gone with the "Cloud" *(energy vehicle)*. The next mention is in Mathew 24; 33A.D. The Olivet discourse. Jesus is upon the Mount of Olives near Jerusalem talking to his disciples about his return at the end of the age and the world situation at that time. "Immediately after the tribulation of those days shall the sun be darkened and the moon shall not give it's light, and the stars shall fall from heaven (meteorites), and the power of the heavens shall be shaken. And then shall appear the sign of the son of man (Jesus Christ) in heaven, and then shall all the tribes of the earth mourn, and they shall see the son of man (Jesus Christ) *coming in the "Clouds of Heaven"* (the same way he left the Earth) with power and great glory. These "Clouds" represent great power, much, much more than anything we have on Earth. The misguided armies of the Earth will attack the returning Jesus. And he (Jesus) will send his Angels (energy beings) with the sound of a "trumpet", and they will gather together his "elect"(human beings chosen from throughout the ages such as the disciples numbering 144,002) from the four winds, from one end of heaven to the other. Reinforced in Paul's letter to the church in Thessalonica (a Greek city) decades later. A.D.54, twenty years after Jesus death and ressurection chapter 3;13 But i would not have you ignorant, brethren, concerning them which are "asleep" (dead) that you sorrow not, the same way others do who have no hope (atheists). For if we believe that Jesus died and rose again, even so, then those also who sleep in Jesus (chosen by Jesus who died), will God bring with him, for this we say unto you by the word of the Lord, that we who remain alive unto the coming of the Lord shall by no means precede those which are asleep (dead). For the Lord himself shall descend from heaven with a shout, with the voice of an Archangel,and with the trump of God, and the dead chosen by Christ shall rise first. Then we who are alive and remain shall be raised up together with them in the "Clouds" (they shall be transported into these "energy vehicles" which already contain the "Angels" and the ressurected "Elect" chosen by Jesus Christ, 144,002 individuals), to meet the Lord in the air (sky), and so shall we always be with the Lord. Living and dead people from all over the world, from many different time periods, being transformed into "spirit" (energy beings-immortals) and being transported up into these energy vehicles "Clouds", where Jesus and his Angels (energy beings) already are in preparation to take over governing the planet just before the

governments of the world destroy it. Jesus left this Earth after his death and ressurection in one of these "Energy vehicles" and is returning with many of them according to these ancient writers.

Mathew 28; Jesus Christ has been crucified and Mary Magdalene goes to his tomb near dawn. "And behold", there was a great eathquake, for an Angel of the Lord descended from heaven(almost certainly in one of these "Clouds"), and came and rolled back the stonefrom the door of the tomb and sat on it. His appearance was like "lightning" (energy) and his clothing white as snow.

And the Angel(energy being) tells the obviously terrified Mary to fear not, Jesus, who was crucified has risen from the dead.

Jesus then appears to his disciples and many others over a period of forty days, and commands his disciples to be witness's about his ministry and ressurection, first in Jerusalem, Judea and Samaria and then to all the world.

And when he had spoken these things, while they looked, he was raised up into the sky and a "Cloud"(energy vehicle) received him out of their sight.

And while they looked up into the sky as Jesus went into the "Cloud"(energy vehicle), behold, two men(energy beings), stood by them in white clothing, who said, men of Galilee, why do you stand looking up into the sky? This same Jesus, who is taken up from you into heaven(in the "Cloud"- energy vehicle), shall come back in the same way as you saw him go into heaven.

Acts chapter 1;

An immortal being, being artificially inseminated into a female human being, (Mary, wife of Joseph), and living out a human existence since birth, experiencing all that human beings experience including extreme pain at his scourging and crucifixion, experiencing human death, and then being ressurected back into immortality. Quite a story, the most amazing story ever told, inadvertantly confirmed by the relationship of these "Clouds of the Lord" throughout mankinds history.

Many atheists throughout history and even today scoff at the suggestion of Mary giving birth as a virgin, with our current technology it's not a big deal, through IVF, So why would it be unreasonable to believe this happened in Jesus time with the technology these beings obviously have. If they can visit earth from somewhere else in the universe it would be a simple matter for them.

"Clouds" visiting Moses, "Clouds" visiting Jesus, "Clouds" taking the ressurected Jesus from the Earth to the ruler of the universe (God), "Clouds" coming back to Earth with Jesus and the Angels to rule the Earth and save mankind from itself, Clouds, Clouds, Clouds, were these ancient people obsessed with "Clouds"? Did they have the intelligence and will to deceive future generations by lying about these " Clouds", or have they just reported what they saw, what they heard, and what they were told by Jesus and these "Energy Beings"?

Throughout all the religious confusion, with thousands of differing doctrines from thousands of differing churches and religions the "Clouds of the Lord" reveal something "extraordinary".

You can be an intelligent, educated person and believe in the fact we are not alone on this planet.

I believe there is a president of the United States of America because i have much tangible evidence of him even though i have never seen him personally.

I also believe in these Energy Beings for many of the same reasons, being the evidence of their existence and their "vehicles" and other knowledge contained in the Bible and other independent sources.

ST MARK

Marks account of Jesus Christ meeting John the baptist, who was Jesus cousin born to Elizabeth(Jesus aunt) in her old age.

And it came to pass, in those days, that Jesus came from Nazareth, in Galilee, and was baptised by John in Jordan. And after coming up out of the water, he saw the heavens opened (a description in many places in the Bible- the sky opening suggesting an entry from another dimension, possibly something similar to a wormhole) and the spirit "like a dove" descending on him. And there came a voice from heaven (the sky), saying, "you are my beloved son", in whom i am well pleased" (a similar description to the "Cloud").

This event hapened around 27 A.D., at the beginning of Jesus Christ's ministry and the end of John the Baptist's.

Mark's account of the transfiguration of Jesus Christ, chapter 9. Near the end of Christ's ministry around 32 A.D.

And after six days Jesus took withhim,Peter,James and John (three of his disciples who ended up being imprisoned and/or killed) and led them up into a high mountain apart by themselves.

And he was transfigured before them, and his raiment became shining, exceeding white as snow, so as no fuller on earth can whiten them, and there appeared unto them Elijah and Moses (both had lived and died many centuries previously) and they were talking with Jesus, and Peter answered and said to Jesus, it is good for us to be here, let us make three tabernacles (tents), one for you, one for Moses and one for Elijah, for he did not know what to say for they were very afraid (understandably so).

And there was a "Cloud" that overshadowed them, and a voice came out of the "Cloud" (the same as it did in Moses time 1400 years previously) saying "this is my beloved son, hear him".

Once again the voice coming from the "Cloud". Have you ever heard this before that a voice came from the "Cloud"? No? The significance of this "Cloud" has been missed by the world, for it is not a raincloud but an "energy vehicle", far, far more superior than our helicopters, planes, spaceships or anything else we have on Earth.

And suddenly when they looked around they saw no man except Jesus, (the "Cloud" gone with Moses And Elijah). Similar descriptions are in Mathew 17 and Luke 9 but not exactly the same showing no collusion.

Mark's account of the Olivet discourse on the mount of olives outside Jerusalem, 33 A.D.

Jesus Christ telling his disciples about conditions in the world prior to his second coming.

And then shall they see the son of man(Jesus Christ) coming in the "Clouds of Heaven" with great power and glory. And then shall he send his Angels(energy beings), and he shall gather together his elect(those chosen by Jesus to rule the planet under him who have been selected throughout the ages), from the four winds, from the uttermost parts of the earth to the uttermost parts of heaven.

People being gathered by these "energy beings" on Jesus Christ's behalf from all over the Earth.

Other parts of the Bible revealing even those chosen who have been long dead being resurrected and beamed up or raised up into these "Clouds" and being transformed into immortals the same as the "Angels", not needing to breathe or eat and being impervious to heat. Can this be accurate and true?

Then let's look at Mark's account of Jesus Christ's arrest just before his crucifixion later in 33 A.D..

With Jesus appearing before the Jewish high priest and chief priests (after judas's betrayal for 30 pieces of silver). The high priest asks Jesus Christ, are you the christ? (the messiah) the son of the blessed?

And Jesus replied I am, and you shall see the son of man (himself) sitting on the right hand of power (God) and coming in the "CLOUDS OF HEAVEN".

Upon hearing this the high priest demanded his execution, and sends him to Pontius Polite(the roman governor) because the execution has to be approved by the ruling roman authorities.

Jesus talking about his future return in these "Clouds of heaven" while he is on trial for his life.

Jesus talking about coming with power in these "Clouds" just before his crucifixion.

These "Clouds" represent much power, as described in the Psalms, "Gods strength is in his "Clouds".

The "Clouds" that the ancient Israelites witnessed in the days of Moses, Solomon,Elijah etc. have had a profound effect on all its witness's.

A proper understanding of these "Clouds", "Clouds of heaven","Clouds of the Lord", "Chariots of fire", "Lights from heaven" etc helps to put these ancient writings into a much easier to understand perspective and allows the reader to see the Bible passages are not fables. The apostle Peter said A.D 66 2-PETER; We have not followed cunningly devised fables (30 years after Christs death and resurrection).

In other words he says this is all real, not a "fairy tale".

By studying the interactions of these "Clouds", "Chariots", "Lights from Heaven etc, it becomes easy to realise that these ancient writers have inadvertantly revealed mankinds involvement with superior beings and far superior technology.

What they have told these ancient writers is even more astounding, combined with what the being referred to as Jesus Christ is reported to have said in the scriptures(the immortal who lived with us as a mortal 2000 years ago, the real story written in these scriptures is quite different from the general story normally preached in churches (which do sound like fables).

Ever heard Jesus Christ is returning to Earth in a "energy vehicle" (Cloud of the lord)? So obvious once you know, once you know how to analyse the evidence properly. No need to rely on university professors who themselves have been indoctrinated into a false belief system of what the Bible means and says.

The "self declared wise" who have stated the Bible is a collection of "fables" written by primitive, superstitious men have revealed themselves to be quite foolish as these scriptures say, "the fool says in his heart there is no God" and " by declaring themselves wise they have become fools".

Anyone can make a mistake but it takes a mature person to admit it.

Just like Albert Einstein, considered to be one of the most intelligent persons to have ever lived, when he saw evidence that the universe had a beginning just like the "primitive" writers of the Bible stated in Genesis 1; (In the beginning, God created the Heavens(universe) and the Earth) he admitted his blunder (which he referred to as the biggest blunder of his career), his "cosmological constant.

Dont make the same blunder, Einstein had the information previously but because the general idea around the "intelligent "elite of the world said the universe was always there he didn't believe the evidence till he found it irrefutable.

One of the last statements recorded of Jesus Christ was his gospel must be preached worldwide in all nations then the end would come (the end of this age and the age of Jesus Christ's return and taking over rulership of the earth with his "elect" (chosen ones).

That preaching to the entire world is now coming to completion with the internet and satellite tv.

By this fact and other related prophecies we know there is not a lot of time left, but all of the Bible prophecies about this matter have to be fulfilled, some are in their final stages, some still need to occur.

WW3 without divine interference would result in the death of all human beings.

Looking at the progression of death and destruction from ww1 to ww2 it is very obvious the level of death and destruction from ww3 would be absolutely devastating on an absolutely horrific scale.

The amount of money being spent worldwide on developing peace as opposed to the amount of money being spent on war capabilities is extremely favourable to war winning out. Only a small percentage is being spent for peace.

The United Nations name would be more accurate if it was called the un-united nations.

Made up of most of the nations on the Earth, including the most powerful, the United States, Russia, China, Britain and France. It hasn't even been able to stop the wars and mass killings in some of the weakest nations on this planet such as Rwanda, Congo and many others.

Because all that the Bible records as having happened past, present and future has occurred, or is occurring, we can be confident that future events recorded in the scriptures will also occur.

http://www.amazon.com/s/ref=nb_sb_noss?url=search-alias%3Ddigital-text&field-keywords=clouds%20of%20the%20lord%20old%20testament

ST MARK

Marks account of Jesus Christ meeting John the baptist, who was Jesus cousin born to Elizabeth(Jesus aunt) in her old age.

And it came to pass, in those days, that Jesus came from Nazareth, in Galilee, and was baptised by John in Jordan. And after coming up out of the water, he saw the heavens opened (a description in many places in the Bible- the sky opening suggesting an entry from another dimension, possibly something similar to a wormhole) and the spirit "like a dove" descending on him. And there came a voice from heaven (the sky), saying, "you are my beloved son", in whom i am well pleased" (a similar description to the "Cloud").

This event hapened around 27 A.D., at the beginning of Jesus Christ's ministry and the end of John the Baptist's.

Mark's account of the transfiguration of Jesus Christ, chapter 9. Near the end of Christ's ministry around 32 A.D.

And after six days Jesus took withhim,Peter,James and John (three of his disciples who ended up being imprisoned and/or killed) and led them up into a high mountain apart by themselves.

And he was transfigured before them, and his raiment became shining, exceeding white as snow, so as no fuller on earth can whiten them, and there appeared unto them Elijah and Moses (both had lived and died many centuries previously) and they were talking with Jesus, and Peter answered and said to Jesus, it is good for us to be here, let us make three tabernacles (tents), one for you, one for Moses and one for Elijah, for he did not know what to say for they were very afraid (understandably so).

And there was a "Cloud" that overshadowed them, and a voice came out of the "Cloud" (the same as it did in Moses time 1400 years previously) saying "this is my beloved son, hear him".

Once again the voice coming from the "Cloud". Have you ever heard this before that a voice came from the "Cloud"? No? The significance of this "Cloud" has been missed by the world, for it is not a raincloud but an "energy vehicle", far, far more superior than our helicopters, planes, spaceships or anything else we have on Earth.

And suddenly when they looked around they saw no man except Jesus, (the "Cloud" gone with Moses And Elijah). Similar descriptions are in Mathew 17 and Luke 9 but not exactly the same showing no collusion.

Mark's account of the Olivet discourse on the mount of olives outside Jerusalem, 33 A.D.

Jesus Christ telling his disciples about conditions in the world prior to his second coming.

And then shall they see the son of man(Jesus Christ) coming in the "Clouds of Heaven" with great power and glory. And then shall he send his Angels(energy beings), and he shall gather together his elect(those chosen by Jesus to rule the planet under him who have been selected throughout the ages), from the four winds, from the uttermost parts of the earth to the uttermost parts of heaven.

People being gathered by these "energy beings" on Jesus Christ's behalf from all over the Earth.

Other parts of the Bible revealing even those chosen who have been long dead being resurrected and beamed up or raised up into these "Clouds" and being transformed into immortals the same as the "Angels", not needing to breathe or eat and being impervious to heat. Can this be accurate and true?

Then let's look at Mark's account of Jesus Christ's arrest just before his crucifixion later in 33 A.D..

With Jesus appearing before the Jewish high priest and chief priests (after judas's betrayal for 30 pieces of silver). The high priest asks Jesus Christ, are you the christ? (the messiah) the son of the blessed?

And Jesus replied I am, and you shall see the son of man (himself) sitting on the right hand of power (God) and coming in the "CLOUDS OF HEAVEN".

Upon hearing this the high priest demanded his execution, and sends him to Pontius Polite(the roman governor) because the execution has to be approved by the ruling roman authorities.

Jesus talking about his future return in these "Clouds of heaven" while he is on trial for his life.

Jesus talking about coming with power in these "Clouds" just before his crucifixion.

These "Clouds" represent much power, as described in the Psalms, "Gods strength is in his "Clouds".

The "Clouds" that the ancient Israelites witnessed in the days of Moses, Solomon, Elijah etc. have had a profound effect on all its witness's.

A proper understanding of these "Clouds", "Clouds of heaven", "Clouds of the Lord", "Chariots of fire", "Lights from heaven" etc helps to put these ancient writings into a much easier to understand perspective and allows the reader to see the Bible passages are not fables. The apostle Peter said A.D 66 2-PETER; We have not followed cunningly devised fables (30 years after Christs death and resurrection).

In other words he says this is all real, not a "fairy tale".

By studying the interactions of these "Clouds", "Chariots", "Lights from Heaven etc, it becomes easy to realise that these ancient writers have inadvertantly revealed mankinds involvement with superior beings and far superior technology.

What they have told these ancient writers is even more astounding, combined with what the being referred to as Jesus Christ is reported to have said in the scriptures(the immortal who lived with us as a mortal 2000 years ago, the real story written in these scriptures is quite different from the general story normally preached in churches (which do sound like fables).

Ever heard Jesus Christ is returning to Earth in a "energy vehicle" (Cloud of the lord)? So obvious once you know, once you know how to analyse the evidence properly. No need to rely on university professors who themselves have been indoctrinated into a false belief system of what the Bible means and says.

The "self declared wise" who have stated the Bible is a collection of "fables" written by primitive, superstitious men have revealed themselves to be quite foolish as these scriptures say, "the fool says in his heart there is no God" and " by declaring themselves wise they have become fools".

Anyone can make a mistake but it takes a mature person to admit it.

Just like Albert Einstein, considered to be one of the most intelligent persons to have ever lived, when he saw evidence that the universe had a beginning just like the "primitive" writers of the Bible stated in Genesis 1; (In the beginning, God created the Heavens(universe) and the Earth) he admitted his blunder (which he referred to as the biggest blunder of his career), his "cosmological constant.

Dont make the same blunder, Einstein had the information previously but because the general idea around the "intelligent "elite of the world said the universe was always there he didn't believe the evidence till he found it irrefutable.

One of the last statements recorded of Jesus Christ was his gospel must be preached worldwide in all nations then the end would come (the end of this age and the age of Jesus Christ's return and taking over rulership of the earth with his "elect" (chosen ones).

That preaching to the entire world is now coming to completion with the internet and satellite tv.

By this fact and other related prophecies we know there is not a lot of time left, but all of the Bible prophecies about this matter have to be fulfilled, some are in their final stages, some still need to occur.

WW3 without divine interference would result in the death of all human beings.

Looking at the progression of death and destruction from ww1 to ww2 it is very obvious the level of death and destruction from ww3 would be absolutely devastating on an absolutely horrific scale.

The amount of money being spent worldwide on developing peace as opposed to the amount of money being spent on war capabilities is extremely favourable to war winning out. Only a small percentage is being spent for peace.

The United Nations name would be more accurate if it was called the un-united nations.

Made up of most of the nations on the Earth, including the most powerful, the United States, Russia, China, Britain and France. It hasn't even been able to stop the wars and mass killings in some of the weakest nations on this planet such as Rwanda, Congo and many others.

Because all that the Bible records as having happened past, present and future has occurred, or is occurring, we can be confident that future events recorded in the scriptures will also occur.

http://www.amazon.com/s/ref=nb_sb_noss?url=search-alias%3Ddigital-text&field-keywords=clouds%20of%20the%20lord%20old%20testament

The Gospel of Luke

There was in the days of Herod, the king of Judea, a certain priest named Zacharias and his wife Elizabeth, and they had no children because Elizabeth was barren, and they were both stricken with age.

And it came to pass when Zacharias was performing his priestly duty in the temple, there appeared to him an Angel of the Lord standing on the right side of the altar.

And when Zacharias saw him he was terrified, but the Angel said to him."fear not", Zacharias, your prayers have been heard, and your wife Elizabeth shall bare a son, and you shall call him John (John the baptist).

Zacharias tells the Angel he and his wife are way too old to bare a child, and the Angel(energy being) says he is Gabriel, who stands before God.

Six months later the being Gabriel goes to a virgin named Mary living in the town of Nazareth, and says, fear not, you will have a child and name him Jesus., and he shall be great and come to rule for ever. Mary says to Gabriel, how can this be? when i have not known a man(had sex).

Gabriel tells her God will inseminate her (artificial insemination for want of a better term) with the "Holy Spirit" and that her elderly cousin Elizabeth is already with child (pregnant) for six months in spite of her old age and that nothing is impossible with God.

There is no mention of where the "Cloud" Gabriel came in is in this chapter but the next chapter mentions "the glory of the Lord" appearing to shepherds,

Luke chapter 2; And there was in the same country, shepherds abiding in the fields, keeping watch over their flock by night, and lo, the Angel of the Lord came upon them, and the "Glory of the Lord" (similar description to Cloud of the Lord) shone round about them, and they were very afraid (undersandably so).

And the Angel said to them, fear not, and tells the shepherds the Saviour has been born in the city of David (Bethlehem).

The shepherds go to Bethlehem and find Mary, Joseph and the babe lying in a manger.

And the shepherds tell everyone what the Angel said, and they call the baby Jesus as Gabriel told them to before the child was conceived.

An old woman (Elizabeth) and a virgin (Mary) conceiving and delivering with the technology Gabriel provided 2000 years ago, we have only had that technology for a few decades, yet these ancient people are reporting it 2000 years ago.

The technology coming from these "Angels" who come here in these "Clouds", and yet some supposedly "wise" people state the Bible and science dont mix.

The Bible, even though it's not a science text book, is full of sound scientific principles.

The story of Adam and Eve, read it yourself in Genesis, I'L give you a scientific translation in modern terms.

Genesis 2-21; And God anaesthetized Adam, and performed an operation on him, and took out one of his ribs, and closed up the flesh again, and the rib God used to "clone" a woman, and Adam called her Eve.

That is what that passage is essentially saying. Nearly 6000 years later we are begining to clone ourselves. The Bible is full of scientific principles, when correctly understood. Don't need an Oxford degree to see that one.

The reason being for the scientific principles in the Bible is these "Angels", (energy beings) have a deep understanding of science.

The sanitation and isolation guidelines in the book of Leviticus have long been considered sound scientific principles well advanced for their time.

Some of these have only been included in our western society in the last 150 years, the ancients were told about them in the time of Moses 3500 hundred years ago.

The practice of these laws has saved millions of people and could have saved many millions more.

Next mention of Jesus Christ is Jesus is about 30 years old.

Luke chapter 3; Now in the fifteenth year of Tiberius Caesar, Pontius Pilate being governor of Judea, and Herod being Tetrarch of Iturea, John the baptist is preaching in the wildernessabout the coming kingdom of God.

Now when all the people were baptised, it came to pass,that Jesus, also being baptised and praying, the heavens were opened, and the holy ghost descended in bodily shape like a dove upon him, and a voice came from heaven (the sky),which said, thou art my beloved son, in whom i am well pleased.

A similar but not exactly the same account as Mathew and Mark showing no collusion to deceive people.

http://www.amazon.com/s/ref=nb_sb_noss?url=search-alias%3Ddigital-text&field-keywords=clouds%20of%20the%20lord%20old%20testament

The Gospel of Luke

There was in the days of Herod, the king of Judea, a certain priest named Zacharias and his wife Elizabeth, and they had no children because Elizabeth was barren, and they were both stricken with age.

And it came to pass when Zacharias was performing his priestly duty in the temple, there appeared to him an Angel of the Lord standing on the right side of the altar.

And when Zacharias saw him he was terrified, but the Angel said to him."fear not", Zacharias, your prayers have been heard, and your wife Elizabeth shall bare a son, and you shall call him John (John the baptist).

Zacharias tells the Angel he and his wife are way too old to bare a child, and the Angel(energy being) says he is Gabriel, who stands before God.

Six months later the being Gabriel goes to a virgin named Mary living in the town of Nazareth, and says, fear not, you will have a child and name him Jesus., and he shall be great and come to rule for ever. Mary says to Gabriel, how can this be? when i have not known a man(had sex).

Gabriel tells her God will inseminate her (artificial insemination for want of a better term) with the "Holy Spirit" and that her elderly cousin Elizabeth is already with child (pregnant) for six months in spite of her old age and that nothing is impossible with God.

There is no mention of where the "Cloud" Gabriel came in is in this chapter but the next chapter mentions "the glory of the Lord" appearing to shepherds,

Luke chapter 2; And there was in the same country, shepherds abiding in the fields, keeping watch over their flock by night, and lo, the Angel of the Lord came upon them, and the "Glory of the Lord" (similar description to Cloud of the Lord) shone round about them, and they were very afraid (undersandably so).

And the Angel said to them, fear not, and tells the shepherds the Saviour has been born in the city of David (Bethlehem).

The shepherds go to Bethlehem and find Mary, Joseph and the babe lying in a manger.

And the shepherds tell everyone what the Angel said, and they call the baby Jesus as Gabriel told them to before the child was conceived.

An old woman (Elizabeth) and a virgin (Mary) conceiving and delivering with the technology Gabriel provided 2000 years ago, we have only had that technology for a few decades, yet these ancient people are reporting it 2000 years ago.

The technology coming from these "Angels" who come here in these "Clouds", and yet some supposedly "wise" people state the Bible and science dont mix.

The Bible, even though it's not a science text book, is full of sound scientific principles.

The story of Adam and Eve, read it yourself in Genesis, I'L give you a scientific translation in modern terms.

Genesis 2-21; And God anaesthetized Adam, and performed an operation on him, and took out one of his ribs, and closed up the flesh again, and the rib God used to "clone" a woman, and Adam called her Eve.

That is what that passage is essentially saying. Nearly 6000 years later we are begining to clone ourselves. The Bible is full of scientific principles, when correctly understood. Don't need an Oxford degree to see that one.

The reason being for the scientific principles in the Bible is these "Angels", (energy beings) have a deep understanding of science.

The sanitation and isolation guidelines in the book of Leviticus have long been considered sound scientific principles well advanced for their time.

Some of these have only been included in our western society in the last 150 years, the ancients were told about them in the time of Moses 3500 hundred years ago.

The practice of these laws has saved millions of people and could have saved many millions more.

Next mention of Jesus Christ is Jesus is about 30 years old.

Luke chapter 3; Now in the fifteenth year of Tiberius Caesar, Pontius Pilate being governor of Judea, and Herod being Tetrarch of Iturea, John the baptist is preaching in the wildernessabout the coming kingdom of God.

Now when all the people were baptised, it came to pass,that Jesus, also being baptised and praying, the heavens were opened, and the holy ghost descended in bodily shape like a dove upon him, and a voice came from heaven (the sky),which said, thou art my beloved son, in whom i am well pleased.

A similar but not exactly the same account as Mathew and Mark showing no collusion to deceive people.

http://www.amazon.com/s/ref=nb_sb_noss?url=search-alias%3Ddigital-text&field-keywords=clouds%20of%20the%20lord%20old%20testament

The Gospel of Luke

There was in the days of Herod, the king of Judea, a certain priest named Zacharias and his wife Elizabeth, and they had no children because Elizabeth was barren, and they were both stricken with age.

And it came to pass when Zacharias was performing his priestly duty in the temple, there appeared to him an Angel of the Lord standing on the right side of the altar.

And when Zacharias saw him he was terrified, but the Angel said to him."fear not", Zacharias, your prayers have been heard, and your wife Elizabeth shall bare a son, and you shall call him John (John the baptist).

Zacharias tells the Angel he and his wife are way too old to bare a child, and the Angel(energy being) says he is Gabriel, who stands before God.

Six months later the being Gabriel goes to a virgin named Mary living in the town of Nazareth, and says, fear not, you will have a child and name him Jesus., and he shall be great and come to rule for ever. Mary says to Gabriel, how can this be? when i have not known a man(had sex).

Gabriel tells her God will inseminate her (artificial insemination for want of a better term) with the "Holy Spirit" and that her elderly cousin Elizabeth is already with child (pregnant) for six months in spite of her old age and that nothing is impossible with God.

There is no mention of where the "Cloud" Gabriel came in is in this chapter but the next chapter mentions "the glory of the Lord" appearing to shepherds,

Luke chapter 2; And there was in the same country, shepherds abiding in the fields, keeping watch over their flock by night, and lo, the Angel of the Lord came upon them, and the "Glory of the Lord" (similar description to Cloud of the Lord) shone round about them, and they were very afraid (undersandably so).

And the Angel said to them, fear not, and tells the shepherds the Saviour has been born in the city of David (Bethlehem).

The shepherds go to Bethlehem and find Mary, Joseph and the babe lying in a manger.

And the shepherds tell everyone what the Angel said, and they call the baby Jesus as Gabriel told them to before the child was conceived.

An old woman (Elizabeth) and a virgin (Mary) conceiving and delivering with the technology Gabriel provided 2000 years ago, we have only had that technology for a few decades, yet these ancient people are reporting it 2000 years ago.

The technology coming from these "Angels" who come here in these "Clouds", and yet some supposedly "wise" people state the Bible and science dont mix.

The Bible, even though it's not a science text book, is full of sound scientific principles.

The story of Adam and Eve, read it yourself in Genesis, I'L give you a scientific translation in modern terms.

Genesis 2-21; And God anaesthetized Adam, and performed an operation on him, and took out one of his ribs, and closed up the flesh again, and the rib God used to "clone" a woman, and Adam called her Eve.

That is what that passage is essentially saying. Nearly 6000 years later we are begining to clone ourselves. The Bible is full of scientific principles, when correctly understood. Don't need an Oxford degree to see that one.

The reason being for the scientific principles in the Bible is these "Angels", (energy beings) have a deep understanding of science.

The sanitation and isolation guidelines in the book of Leviticus have long been considered sound scientific principles well advanced for their time.

Some of these have only been included in our western society in the last 150 years, the ancients were told about them in the time of Moses 3500 hundred years ago.

The practice of these laws has saved millions of people and could have saved many millions more.

Next mention of Jesus Christ is Jesus is about 30 years old.

Luke chapter 3; Now in the fifteenth year of Tiberius Caesar, Pontius Pilate being governor of Judea, and Herod being Tetrarch of Iturea, John the baptist is preaching in the wildernessabout the coming kingdom of God.

Now when all the people were baptised, it came to pass,that Jesus, also being baptised and praying, the heavens were opened, and the holy ghost descended in bodily shape like a dove upon him, and a voice came from heaven (the sky),which said, thou art my beloved son, in whom i am well pleased.

A similar but not exactly the same account as Mathew and Mark showing no collusion to deceive people.

http://www.amazon.com/s/ref=nb_sb_noss?url=search-alias%3Ddigital-text&field-keywords=clouds%20of%20the%20lord%20old%20testament

Acts of the Apostles

After Jesus Christ was crucified and resurrected from the dead, he appears,walks,talks, eats with his disciples for forty days.

At the end of the forty days, Acts 1-8; But you(the apostles) shall receive power, after the holy ghost is come upon you.

You shall be witness's about me in Jerusalem, in Judea and Samaria

and to the uttermost parts of the earth.

And when he had spoken these things, while they looked, he was taken up into the sky, and a "Cloud" received him out of their sight.

The same as the "Clouds" which visited Moses and Elijah 1500 and 1000 years previously, the same "Clouds" which Moses and Elijah came to visit Jesus in not long before his crucifixion, the same "Clouds" which the Israelites followed in the wilderness and passed through the sea under these "Clouds", The same "Clouds" which God visited the finished tabernacle in the desert with Moses, the same "Clouds" which visited the finished temple of Solomon, the same "Clouds" which Jesus said he would return to the Earth in to govern the Earth with Power and Glory, the same "Clouds" mentioned in the book of Psalms,"Gods strength is in his "Clouds", who makes these "Clouds" his Chariots".

These Clouds are not rainclouds, they are energy clouds/vehicles which the immortal/energy beings make their "Chariots".

The ancient writers used these terms to describe what they witnessed, the Angels/energy beings also used these terms so these primitive people could understand somewhat.

These "Clouds" are vehicles used by these beings who the ancient writers refer to as God And Angels.

The next chapters refer to Jesus Christs return to Earth

Acts 1,10; And while they looked steadfastly toward heaven as he went up into the sky, behold, two men stood by them in white apparel(Angels), who said, "you men of Galilee, why stand you gazing up into Heaven, this same Jesus, who is taken up from you into Heaven, shall come back the same way you saw him go into Heaven.

Left Earth after his resurrection in a "Cloud", coming back to Earth in a "Cloud".

This being we refer to as Jesus Christ telling his disciples to witness his resurrection to the whole world.

Which is precisely what has happened, the rest is history. Christianity in spite of all the confusion in the many diverse churches. They all believe in Christs resurrection, Catholic, Protestant, Orthodox etc.

Since we do not have the ability today to resurrect the truly dead, the technology that brought Jesus Christ back to life after being dead for three days two thousand years ago, came from another wordly source, which the ancient writers inadvertantly reveal as coming from the "Angels" who came to Jesus Christs Tomb in a "Cloud".

It is obvious that something spectacular happened two thousand years ago in Jerusalem that has affected the whole world.

Its also obvious that something spectacular happened 3500 years ago in the middle east in Moses time, which still has a profound effect upon the Jews and Muslims and Christians of the world.

Today,thousands of years later with the help of scientific reasoning and applying unbiased logic, we can determine what happened much more clearly than those who were alive during the dark and middle ages, and we dont have to rely on the educated "elite", we can investigate ourselves, the educated elite of the middle ages hid the knowledge in the Bible by keeping it in the language of the elite,Latin, translating it into another language, or even posession of a Bible which has been translated(or any piece of scripture not written in latin) would likely result in horrific torture and being burnt at the stake (read the story of Jan Hus http://en.wikipedia.org/wiki/Jan_Hus).

Many of the educated elite of today are also misleading the masses, ridiculing the Bible, God, and anyone who does not hold onto their unrealistic views.

Many people died horrible deaths to allow everyone access to this book, so why not see if what they died for has something to offer you

(besides a horrible death at the stake).

Clouds of the Lord helps you to understand what has happened in the past, give you some understanding of the present, and finally to understand the future. Clouds of the Lord is a key to unlocking the mysteries in the Bible.

It's not the only key but it is a very important one.

http://www.amazon.com/s/ref=nb_sb_noss?url=search-alias%3Ddigital-text&field-keywords=clouds%20of%20the%20lord%20old%20testament

Acts of the Apostles

After Jesus Christ was crucified and resurrected from the dead, he appears, walks, talks, eats with his disciples for forty days.

At the end of the forty days, Acts 1-8; But you(the apostles) shall receive power, after the holy ghost is come upon you.

You shall be witness's about me in Jerusalem, in Judea and Samaria

and to the uttermost parts of the earth.

And when he had spoken these things, while they looked, he was taken up into the sky, and a "Cloud" received him out of their sight.

The same as the "Clouds" which visited Moses and Elijah 1500 and 1000 years previously, the same "Clouds" which Moses and Elijah came to visit Jesus in not long before his crucifixion, the same "Clouds" which the Israelites followed in the wilderness and passed through the sea under these "Clouds", The same "Clouds" which God visited the finished tabernacle in the desert with Moses, the same "Clouds" which visited the finished temple of Solomon, the same "Clouds" which Jesus said he would return to the Earth in to govern the Earth with Power and Glory, the same "Clouds" mentioned in the book of Psalms,"Gods strength is in his "Clouds", who makes these "Clouds" his Chariots".

These Clouds are not rainclouds, they are energy clouds/vehicles which the immortal/energy beings make their "Chariots".

The ancient writers used these terms to describe what they witnessed, the Angels/energy beings also used these terms so these primitive people could understand somewhat.

These "Clouds" are vehicles used by these beings who the ancient writers refer to as God And Angels.

The next chapters refer to Jesus Christs return to Earth

Acts 1,10; And while they looked steadfastly toward heaven as he went up into the sky, behold, two men stood by them in white apparel(Angels), who said, "you men of Galilee, why stand you gazing up into Heaven, this same Jesus, who is taken up from you into Heaven, shall come back the same way you saw him go into Heaven.

Left Earth after his resurrection in a "Cloud", coming back to Earth in a "Cloud".

This being we refer to as Jesus Christ telling his disciples to witness his resurrection to the whole world.

Which is precisely what has happened, the rest is history. Christianity in spite of all the confusion in the many diverse churches. They all believe in Christs resurrection, Catholic, Protestant, Orthodox etc.

Since we do not have the ability today to resurrect the truly dead, the technology that brought Jesus Christ back to life after being dead for three days two thousand years ago, came from another wordly source, which the ancient writers inadvertantly reveal as coming from the "Angels" who came to Jesus Christs Tomb in a "Cloud".

It is obvious that something spectacular happened two thousand years ago in Jerusalem that has affected the whole world.

Its also obvious that something spectacular happened 3500 years ago in the middle east in Moses time, which still has a profound effect upon the Jews and Muslims and Christians of the world.

Today,thousands of years later with the help of scientific reasoning and applying unbiased logic, we can determine what happened much more clearly than those who were alive during the dark and middle ages, and we dont have to rely on the educated "elite", we can investigate ourselves, the educated elite of the middle ages hid the knowledge in the Bible by keeping it in the language of the elite,Latin, translating it into another language, or even posession of a Bible which has been translated(or any piece of scripture not written in latin) would likely result in horrific torture and being burnt at the stake (read the story of Jan Hus http://en.wikipedia.org/wiki/Jan_Hus).

Many of the educated elite of today are also misleading the masses, ridiculing the Bible, God, and anyone who does not hold onto their unrealistic views.

Many people died horrible deaths to allow everyone access to this book, so why not see if what they died for has something to offer you

(besides a horrible death at the stake).

Clouds of the Lord helps you to understand what has happened in the past, give you some understanding of the present, and finally to understand the future. Clouds of the Lord is a key to unlocking the mysteries in the Bible.

It's not the only key but it is a very important one.

http://www.amazon.com/s/ref=nb_sb_noss?url=search-alias%3Ddigital-text&field-keywords=clouds%20of%20the%20lord%20old%20testament

The CONVERSION of SAUL/PAUL

Chapter 9; And Paul, talking about beating and slaughtering the disciples of Jesus (after his Crucifixion) went to the Jewish high priest, and asked him for letters to take to Damascus(Syria) synagogues to get help arresting the followers of Jesus and bring them to Jersusalem for judgement, he was a fanatical jew, willing to even kill for his religous beliefs.

And as he journeyed, he came near damascus, and suddenly there shone round about him, a "light from heaven", and he panicked, and he heard a voice say to him, Saul, Saul, why do you persecute me? and Saul says "who are you lord"? I am Jesus replies the voice.

The rest is history, this Jewish fanatic became one of the staunchest christians of all time.

A "light from heaven", A "Cloud of the Lord", "Chariots of God", or my personal description " Energy Vehicle". I believe none of these descriptions to be satisfactorarily accurate.

Nearly thirty years later Paul recounts the events that changed him from a christian persecuting fanatic to a staunch supporter of Jesus Christ, after being arrested and brought to account before the king, Agrippa and the roman governor.

Acts 26; at midday o'king, i saw in the way "a light from heaven", above the brightness of the Sun shining round about us.

And then recounts Jesus Christ talking to him.

Pauls letter to the church in Corinth, chapter 10; more over brethren, i dont want you to be ignorantof the fact that all our fathers were under the "Cloud" (energy vehicle), and they all passed through the divided sea on dry land, with the waters standing up on their left and right sides (a force field from the "Cloud" (energy vehicle) holding the waters back for the Israelites to cross). And they were all baptised with Moses in the "Cloud" and in the sea. (the Egyptians drowning behind them in the collapsing waters as the force field is removed).

Paul inadvertantly revealing the "Cloud" parting the sea with Moses and the Israelites crossing through the sea under the "Cloud".

The "Cloud" not mentioned in this particular act in the Old Testament makes the Old Testament account more understandable.

THE ENERGY VEHICLE/CLOUD PARTING THE SEA AND USING A FORCEFIELD TO HOLD BACK THE WATER FOR THE iSRAELITES TO CROSS UNDERNEATH THE "CLOUD"./ENERGY VEHICLE.

Paul's letter to the church in Thessalonica;

But i dont want you to be ignorant, brethren, concerning them which are asleep(dead), that you have hope, for if we believe that Jesus died and rose again, then those who sleep(are dead) in Jesus, God will bring with him.

For the Lord himself shall descend from Heaven with a shout, and the dead in Christ(the elect-chosen ones) shall rise first.

Then we which are alive at that time shall be raised up together with those already in the "Clouds" (energy vehicles) and we shall stay with the Lord.

When Jesus Christ was alive he said he would return in these "Clouds" of Heaven with power and glory.

When Jesus ascended to heaven he went into one of these "Clouds".

While the Apostles watched Jesus going up into the sky into the "Cloud", two Angels told them he would come back the same way (in a "Cloud").

Living and dead people being raised up (or "beamed " up) into these "Clouds" of heaven where the returning Jesus Christ and his Angels are.

http://www.amazon.com/s/ref=nb_sb_noss?url=search-alias%3Ddigital-text&field-keywords=clouds%20of%20the%20lord%20old%20testament

The book of revelation-The last book of the Bible .

Written byJohn, Jesus Christ's last living apostle around 96A.D.

John has been imprisoned on the isle of Patmos by the Roman authorities for his preaching.

It Begins with, the revelation of (the resurrected) Jesus Christ which God gave to him to show his servants things which must happen,and he sent it to his disciple John.

John sends the revelations to the seven churches in Asia.

From Jesus Christ, who is the first to be raised from the dead, who is immortal, who is to return to Earth

Behold he comes with "Clouds" and everyone shall see him .

John was commanded to write down the things he has seen, the things which are (the present), and the things to come (the future).

John goes onto write about a future period in time, which appears to sound much like today. with unusual seismic activity, weather conditions and wars, culminating in one third of the entire Earths population being destroyed.

chapter 9-14.

Saying to the sixth Angel, loose the four Angels which are bound in the Euphrates(the main river in Iraq),

and goes on to describing massive armies involved in warfare destroying the third part of mankind.

The Iraq war has set off events that have made the world a much more volatile place increasing islamic extremism and anti-west attitudes across the globe.

The next chapter - And I saw another mighty "Angel" come down from Heaven ,clothed with a "Cloud" (once again coming from space in a "Cloud"), and a rainbow was upon his head, and his face was as it were the sun, and his feet as pillars of fire (desciption of an energy being). A description of future events which John saw in vision, apparently beamed into Johns eyes or mind, like a movie.

Technology mankind is only in its infancy developing today. (vision technology)

Then in chapter 11 John hears.

And I will give power to my two Witness's (Witness's of God and Jesus Christ), and they shall prophesy one thousand,two hundred and sixty days.

They shall then be killed in Jerusalem at the end of this period, their dead bodies lying in the street, for three and a half days, after this time the two Witness's are resurrected ,and a voice from heaven says to them to come up here, and they ascended up to heaven (the sky) into a "Cloud" (energy vehicle) this happens when Jesus Christ returns to the Earth with his "Angels" in "the Clouds of heaven" and the elect(chosen ones of Jesus Christ) are also brought into the returning "Clouds". In line with a similar account in Paul's letter to the Thessalonians about Christs return.

Then we which are alive and remain shall be brought up (beamed up) to them already in the "Cloud".(The resurrected dead ,Jesus Christ and his Angels) in the sky.

The ancient writers inadvertantly revealing interactions with beings who have fantastic technology, Who claim they can resurrect the dead, even if their bodies have decomposed completely,and make them immortal with new bodies, which time and the elements have no effect on.

I would imagine if you have read this book and studied and checked to see if it is true, you would make the logical assumption that mankind has had interactions with beings of much higher intelligence than ours.

So who are they? Aliens from another planet?, Humans from the future? Who do they say they are?

The gospel (good news) according to John .

In the beginning was the Word and the Word was with God.(the Word is a title for Jesus Christ).

All things were made by him .A statement that says Jesus was involved in creating the universe.

The first chapter of Genesis- In the beginning God created the heavens and the earth.

The sixth day- God said let us make man in our own image,and let him rule over the whole earth.

Chapter 2- And the lord God formed a man out of the dust of the earth(most likely genetic material such as bone to make our current species) and breathed into him the breath of life.Then God placed the man in a garden called Eden, in modern day Iraq. As our creators they refer to themselves as our Father and his Son, Our older and much wiser brother, offering us to join them as immortals, if we will accept them as such and follow their guidance. Thats the primary message in the Gospel(good news). We have the offer of immortality from these beings. As I said in the forward of "Clouds of the Lord" old testament.

The "Clouds of the Lord" are only one area of the consistency of the Bible, many so called "bible scholars" and "preachers" have little idea or understanding of the Bible. Here you can check the information posted and see if it is correct. Do not blindly believe the supposed "experts". The Bible itself warns against these so called "wise men" and describes them as "fools"..

Another statement in the Bible- I am God and there is none like me, declaring the end from the beginning, and from ancient times things not yet done. The Bible being in existence today is no "amazing coincidence", it was declared from ancient times to reach us today, to give us guidance and hope.

The Bible is full of statements which are similar and for the scoffers, particularly the ones who write these misleading books, stating that the Bible is a collection of fables by primitive man and that anyone who believes it is backward, this one in particular,

The fool says in his heart 'there is no God', And by professing themselves to be wise(eg:by writing these misleading books) they have become fools.

Not me saying this,but God, thousands of years ago. (if you believe this book) or if you believe the writers of these misleading books, by primitive man writing "fables".

My advice? Make up your own mind. These misleading books actually help to confirm the authenticity of the Bible, "In the last days, scoffers will come, saying where is he?(Jesus Christ) things go on like they always have .

So if there were no scoffers then that statement would not be accurate.

These misleading writers inadvertantly help to authenticate the Bible, It can be quite hard to believe the Bible without the correct information, and the part of the "Clouds of the Lord", but a thorough investigation reveals time and again the same answer. Interaction with supernatural (or super intelligent beings). Who want us to ultimately join them as immortals. Perhaps the Bible really is correct, and if it is it is certainly worth investigating personally. Your choice, investigate yourself. You do not need to have harvard or oxford degrees for this, when I was at school in the sixties and seventies I was told that the universe was endless and has always existed.

Albert Einstein (one of the greatest physicists of all time) believed this to be so in spite of his calculations and genius.

When he found out the Universe had a beginning ,and was not static, he called it (his cosmological constant) the biggest blunder of his career.

We now know the universe had a beginning about fourteen billion years ago(the big bang) and that everything is travelling through the universe at enormous speed,the earth is travelling through space around the sun at 67 thousand miles per hour.

,

Everything is moving at terrific speed in all directions from a central point from where the big bang occurred approximately 14billion years ago. Imagine a balloon being blown up to enormous size so that it inflates continually and everything in the universe is contained within this balloon.

The first chapter in Genesis, in the beginning God created the heavens (universe) and the Earth.

The Bible stating thousands of years ago that the Universe had a beginning, whereas the intellectual elite of the 19 and early 20th century said the universe had no beginning, it was always there, even Albert Einstein fell for this deception, Calling it the biggest blunder of his career.

Whats at this central point where the universe started? where the balloon was blown up from? - Book of Revelation-20-11.

And I saw a great white throne,and him that sat on it, from whom the Earth and the Heavens fled away.

The Earth and the Universe moving quickly away from God.which is to be expected from the creator of the Universe whose throne is at the centre of the entire universe(yet outside the universe, imagine a person blowing up a gigantic balloon, how everything would expand from that beginning point.) where he created it from 14 billion odd years ago.

There is enormous evidence of design throughout the Earth and entire Universe, understanding the role of the "Clouds of the lord" helps to understand some of the genius of the designers of Earth.

Beings who reveal to us that all creation is our ultimate inheritence for all eternity.

We will not be allowed to conquer the universe till we conquer ourselves (conquer our defects of character).

http://www.amazon.com/s/ref=nb_sb_noss?url=search-alias%3Ddigital-text&field-keywords=clouds%20of%20the%20lord%20old%20testament

www.ingramcontent.com/pod-product-compliance
Lightning Source LLC
Chambersburg PA
CBHW081853170526
45167CB00007B/3002